Food 116

虫 草

Insect Grass

Gunter Pauli

冈特·鲍利 著

凯瑟琳娜·巴赫 绘
颜莹莹 译

丛书编委会

主　任：贾　峰

副主任：何家振　闫世东　郑立明

委　员：牛玲娟　李原原　李曙东　李鹏辉　吴建民
　　　　彭　勇　冯　缨　靳增江

特别感谢以下热心人士对译稿润色工作的支持：

王必斗　王明远　王云斋　徐小怗　梅益凤　田荣义
乔　旭　张跃跃　王　征　厉　云　戴　虹　王　逊
李　璐　张兆旭　叶大伟　于　辉　李　雪　刘彦鑫
刘晋邑　乌　佳　潘　旭　白永喆　朱　廷　刘庭秀
朱　溪　魏辅文　唐亚飞　张海鹏　刘　在　张敬尧
邱俊松　程　超　孙鑫晶　朱　青　赵　锋　胡　玮
丁　蓓　张朝鑫　史　苗　陈来秀　冯　朴　何　明
郭昌奉　王　强　杨永玉　余　刚　姚志彬　兰　兵
廖　莹　张先斌

目录

虫草	4
你知道吗？	22
想一想	26
自己动手！	27
学科知识	28
情感智慧	29
艺术	29
思维拓展	30
动手能力	30
故事灵感来自	31

Contents

Insect Grass	4
Did you know?	22
Think about it	26
Do it yourself!	27
Academic Knowledge	28
Emotional Intelligence	29
The Arts	29
Systems: Making the Connections	30
Capacity to Implement	30
This fable is inspired by	31

一只毛虫正在寻找掩护。她试图快速蠕动,却被灌木丛绊住了。一只渡鸦大笑着,看着他的午餐笨拙地想要逃跑。

"你为什么这么匆忙?"渡鸦问。

"因为我闻到了空气中蘑菇的味道。"毛虫颤抖着声音说。

A caterpillar is heading for cover. She tries to move fast, but trips and stumbles, and becomes tangled in the undergrowth. A raven laughs at the sight of his lunch clumsily trying to escape.

"Why are you rushing away like this?" Raven asks.

"Because I smell mushrooms in the air," Caterpillar replies in a trembling voice.

一只毛虫正在寻找掩护……

A caterpillar is heading for cover ...

为了躲避蘑菇而逃跑？

Running away from mushrooms?

"为了躲避蘑菇而逃跑？别开玩笑了，你应该躲的是我！"

"这些蘑菇很危险。一旦沾上我的皮肤，它们就会慢慢地占领我的整个身体，最后杀死我。"

"Running away from mushrooms? You must be joking. It is me you should run away from!"

"These mushrooms are very dangerous. Once attached to my skin, they will slowly take over my whole body and kill me."

"是啊,你太容易被发现了。而且你又大又笨拙。但是我还不知道蘑菇能够从空中袭击你呢。"

"这些蘑菇就像隐形战斗机。它们长在泥土里,看上去像普通的菌类,但是只要哪怕一丁点儿落在我身上,我就得为自己准备葬礼了。"

"哇,那可真是可怕的蘑菇,竟然能够导致慢性死亡!它有名字吗?"

"Well, you are very easy to spot. And you are big and clumsy. But I didn't know mushrooms are able to fly around and attack you from the air."

"These mushrooms are like stealth fighters. They settle in the soil, looking just like a mould, but once even just a tiny bit of it gets trapped on me, I may as well prepare for my funeral."

"Wow, that is a scary mushroom! One that causes a slow death. Does it have a name?"

这些蘑菇就像隐形战斗机……

These mushrooms are like stealth fighters ...

它叫"虫草"

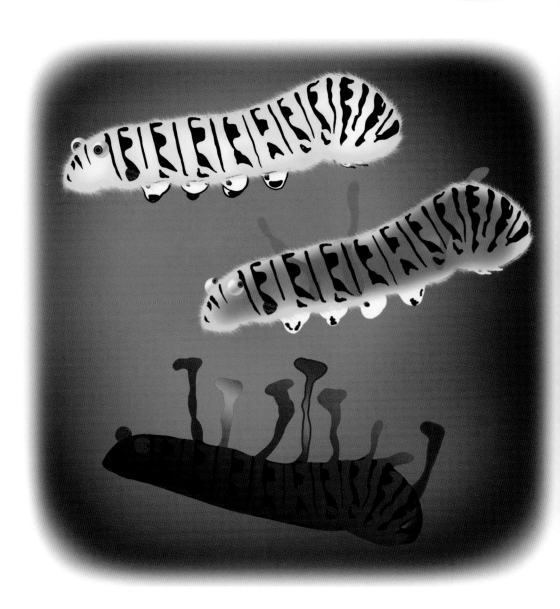

It is called "insect grass"

"它叫'虫草'。"毛虫答道。

"真的吗?为什么叫'虫草'呢?是不是意味着在你死后你的肚子里就会长出草来?至少这样有益于形成养分丰富的表层土。你知道,我们现在正缺健康的表层土。"

"嗯,其实从我身上长出来的并不是真的草,而是菌类。"

"It is called 'insect grass'," Caterpillar replies.

"Really, why 'insect grass'? Does that mean that grass will grow on your tummy after you die? At least that will be good for healthy topsoil. You know we cannot get enough of that."

"Look, the 'grass' that will grow on me is not really grass, it is a mushroom."

"人类什么时候才能学会使用正确的名称来描述事物呢？还是他们就是想让人感到困惑？"渡鸦问道。

"这个特别的蘑菇才不会让他们困惑呢。他们喜欢得不得了——愿意不惜重金得到它，然后吃掉。"

"When will people finally learn to use the correct names for things? Or are they out to confuse everyone?" Raven asks.

"When it comes to this particular mushroom, people are not confused at all. They adore it – and are ready to pay any price to have it and eat it."

愿意不惜重金得到它，然后吃掉……

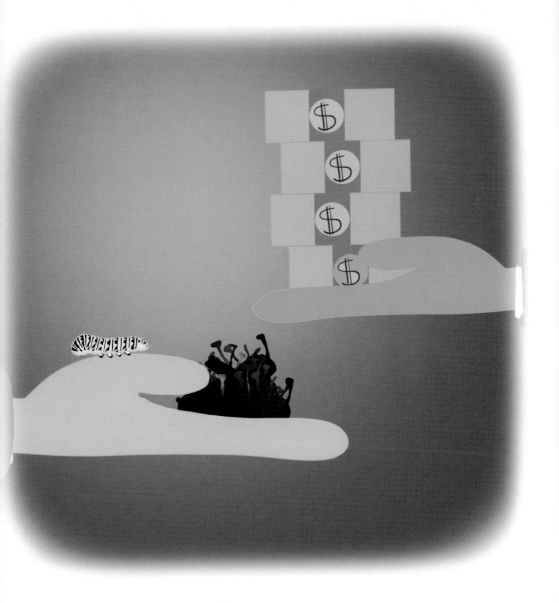

Will pay any price to have and eat it …

将我制成某种特殊药材！

Use me to make special medicine!

"竟然这样！人们喜欢从毛虫身上长出来的'草'？每天我都能了解到有关人类的奇怪又新鲜的事情。"

"说实话，我并不介意作为生命网络的一部分。既然就像地球上其他生命一样，我终将归于尘土，干吗不将我制成某种特殊药材呢？"

"Really! People love 'grass' that grows on a caterpillar's body? I am learning strange new things about the human race every day."

"To be honest, I do not mind that I am part of the web of life. As – like anyone else on Earth – I am returning to dust anyway, why not use me to make some special medicine?"

"你想多了吧！从你肚子里长出来的草被用作药材？真的吗？"

"当然。"

"说来听听。"

"Now you are taking your fantasy a bit too far! The grass growing in your tummy used as medicine, really?"

"You bet."

"Then tell me more."

药材？真的吗？

Medicine, really?

抓到我当午餐……

Snatch me up for your lunch ...

"嗯，或许你只是假装感兴趣，让我一直说话，这样你就可以抓到我当午餐了吧。在你飞行时从你身体里排出的东西又不能制成药材。"

"这就是你给我的不让我吃掉你的最好理由了吗？"渡鸦问。

"Well, you may just pretend to be interested, to keep me talking, so you can snatch me for your lunch. And no medicine can be produced from what exits from the other side of you during your flight."

"Is that the best reason you can give me not to eat you?" Raven wants to know.

"你看,我们俩都不是科学家,但是东西方的医生们一致认为这些菌类富含营养成分,是自然界的一个奇迹!"

"所以人类想给它加冕,让它成为真菌王国的新任女王吗?"

"不,不,还没有,但是很显然,一颗新星诞生了。"

……这仅仅是开始!……

"Look, I am not a scientist, and neither are you, but doctors from the East and the West agree that these mushrooms are so full of good things, that they consider it a marvel of Nature!"

"So do people now want to crown it the new Queen of the Fungus Kingdom?"

"No, no! Not yet, anyway – but it is clear that a star is born."

... AND IT HAS ONLY JUST BEGUN!...

……这仅仅是开始!……

...AND IT HAS ONLY JUST BEGUN!...

Did You Know?
你知道吗？

Ravens are wild, wary, bold, smart, suspicious and social birds. They love to fly and do acrobatic stunts with tricks like twisting, turning, side slipping, looping and nose-diving.

渡鸦是野蛮、谨慎、大胆、聪明、机警和世故的鸟类。它们酷爱飞行时做杂技表演：盘旋、转身、滑行、环绕及俯冲。

Ravens are considered as intelligent as chimpanzees and dolphins. During preening these birds practice "anting" by chewing ants and rubbing their guts on their feathers, to act as an insecticide and fungicide. Ravens also make toys to play with, alone or as a team.

渡鸦与黑猩猩和海豚一样聪明。这种鸟会用"蚂蚁浴"整理羽毛，它们咀嚼蚂蚁并将蚂蚁的内脏擦在羽毛上，杀虫除菌。渡鸦还会制作玩具，有时单独玩，有时群体一起玩。

Shakespeare presented the raven as messengers of evil in plays such as *Julius Caesar*, *Macbeth* and *Othello*, while in *Titus Andronicus* he describes ravens as benefactors who feed abandoned children.

在莎士比亚的戏剧诸如《尤利乌斯·恺撒》《麦克白》和《奥赛罗》中,渡鸦是邪恶的信使,但是在《泰特斯·安特洛尼克斯》里,渡鸦被描绘成喂养弃婴的恩人。

The cordyceps mushroom is popular under the name caterpillar fungus or insect grass. In Tibet, it is called "winter worm, summer grass" since it invades the host in the fall, and pops out as if it were a grass in the summer.

虫草菌为人熟知的名称是虫草。在西藏,它被称为"冬虫夏草",因为它秋天的时候侵入幼虫体内,到夏天的时候像草一样破土而出。

The mushroom cannot reproduce if all infected hosts are harvested. In Bhutan, the farmers will leave about one-third of the insects behind, worth thousands of dollars, in order to ensure a harvest for the following year.

如果将所有受到感染的虫体都采集走，虫草菌就不能繁殖了。在不丹，农户为保证下一年的收成，会保留约三分之一的幼虫，价值数千美元。

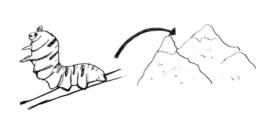

The caterpillars prone to be infected by this fungus, live in the mountain plateaus at an altitude of 3000 to 5000 metres. The fungus consumes its host as it hibernates.

蝠蛾科幼虫容易被海拔3000—5000米高原地区的虫草菌感染。虫草菌会在幼虫过冬时吸取虫体中的养分。

Traditional Chinese Medicine (TCM) calls cordyceps an 'insect plant' and considers the fungus to be an excellent balance of yin and yang as it is composed of both animal and plant (fungus). It is also considered to have the properties of sweetness and warmth.

传统中医称冬虫夏草为虫草，认为虫草能够很好地平衡阴阳，因为它是动物与植物（真菌）的复合体。中医认为，虫草味甘，性温。

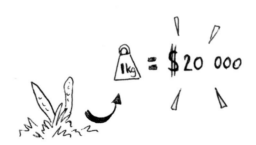

Households in rural areas where cordyceps is found, depend on harvesting this fungus for nearly half of their income. The price for one kilogram of high quality (big size) cordyceps can reach US$ 20 000 in the village and US$ 100 000 in Chinese megacities.

在偏远地区靠采集虫草为生的家庭，近一半的收入来自虫草。一公斤优质（虫体较大）的虫草在农村能卖到 20 000 美元，在中国的大城市卖到 100 000 美元。

Think About It 想一想

Does it sound appetising to eat a fungus growing on a caterpillar?

去吃长在毛虫身上的菌类，这听上去令人有食欲吗？

What do you think of a bird that can talk and make its own toys to play with?

你如何评价一只既能说话又能给自己做玩具的鸟？

If it was up to you, would you harvest all of the very precious fungi, or would you leave some for the season?

你会把所有珍贵的菌类全都采集走，还是会留下一些？

Is Nature a good source of medicine?

大自然是一个很好的药物来源吗？

The raven plays a symbolic role in the imagination of diverse cultures throughout history. Make an inventory of all the attributes that ravens have acquired, in cultures ranging from the Native Americans to the Celts, Vikings, Irish, English, Bhutanese, Chinese and Tibetans. Now compare the positive and the negative characteristics, before forming your own opinion. What would you like the raven to symbolise for you? Explain your reasoning to your friends and family.

渡鸦在历史上不同文化的想象中起着象征作用。记录一下渡鸦在不同文化中表现出的所有象征意义，包括美洲原住民、凯尔特人、维京人、爱尔兰人、英格兰人、不丹人及中国人。比较一下这些正面和负面特征，然后得出你的结论。对你来说渡鸦象征什么呢？向朋友及家人说一下你的理由。

TEACHER AND PARENT GUIDE

学科知识
Academic Knowledge

生物学	渡鸦与乌鸦的区别；渡鸦是杂食动物，主要以肉类、啮齿动物和蛋类为食；虫草属于子囊菌门，约有1500个品种；虫草是昆虫与真菌的寄生体；真菌的菌丝趁宿主冬眠时侵入其体内；虫草能够影响宿主的行为；动植物和真菌残体通过微生物作用转变成土壤。
化 学	由某几种虫草中提炼出的虫草素中含有环孢菌素，是一种免疫抑制剂，可以抑制器官移植的排异反应；虫草能够补充芬戈莫德（一种脂类）从而治疗多发性硬化症；虫草中的多糖和虫草素具有抗癌活性；虫草含有蛋白质、多肽、氨基酸、多胺、糖类以及固醇、脂肪酸，维生素B_1、B_2、B_{12}、E、K。
物 理	虫草的孢子粘在毛虫的毛细孔上，进入其体内，这一过程只需考虑其大小及表面张力，并不需要发生生物或化学反应。
工程学	虫草被应用于新兴的生物制药工程；虫草的生长依靠温度、土壤压力及光强度；潜行：使雷达无法探测到行动的能力。
经济学	虫草是一种传统中药材，售价很高，也是器官移植成功的关键；从野生虫草中移植出一部分虫草，用来大规模工业化种植，价格也有所降低；在喜马拉雅地区，当地居民采集虫草得来的收入占其总收入的一半。
伦理学	如果把所有的东西都收割了，明年就没有收成，如果有节制地采集虫草，留一部分继续生长，就能够可持续地收获；我们是否应该以破坏农村经济、使农民丧失生计为代价，来发展工业呢？
历 史	1885年，通萨的地方首领晋美朗杰（不丹第一任国王的父亲）与英国人打仗时带着渡鸦王冠作为头盔，所以渡鸦王冠成为以后所有国王的盔甲；埃德加·爱伦·坡在1845年以他的一首诗《渡鸦》使这种鸟名垂后世。
地 理	渡鸦是不丹的国鸟，不丹是虫草的主要产地之一；渡鸦在西藏、锡金、拉达克和不丹有着很高的地位。
数 学	一公斤小麦种子能产八公斤用来做面包的小麦；指数函数：如果一粒大米乘以二，之后每天大米的数量都是前一天的两倍。
生活方式	为将来而做储备的必要性；人们存得越来越少，借得越来越多，把未来的收入用于还贷，而不是按照自己的意愿生活；保存两倍于实际需要的积蓄，这样万一遇到困难，还有足够的储备，一年后从头再来。
社会学	不丹的主要守护神之一曾以渡鸦的形象现身；"渡鸦青藏亚种"这一学名得自西藏，但不丹也是渡鸦的故乡；渡鸦总是与死亡和厄运相关联；凯尔特战争女神在战斗时会化身为渡鸦；维京神奥丁有两只大乌鸦，Hugin（思想）和 Munin（记忆），它们每天飞到世界各地，每晚向奥丁汇报当天的见闻。
心理学	狂想状态；乡村女孩智斗自私的国王，向其要求在30天内每天分给她的米都是前一天的2倍，以此解救饥饿的人们；为得到更大利益而做出牺牲。
系统论	过度收割会破坏下一年的收成；生命网络。

教师与家长指南

情感智慧
Emotional Intelligence

渡 鸦

渡鸦很率直,他本可以立刻就吃掉毛虫,但还是问了毛虫一个显而易见的问题。渡鸦并没把蘑菇放在眼里,他认为对于毛虫来说他才是更大的威胁。他出言不逊,但后来意识到他对蘑菇的危险性缺乏认知。他对蘑菇的名称感到困惑,因为这个名称不能说明它实际是什么。渡鸦认为错误的名称会导致迷惑。他承认他在了解新的事物,但很快又认为是毛虫陷入幻想。当毛虫固执己见时,渡鸦让毛虫更深入地解释原因。这对毛虫来说是个敏感的话题,因为它知道这只渡鸦想以玩笑轻松地结束这场对话。

毛 虫

毛虫有点偏执。渡鸦对她产生了同理心,毛虫便更详细地向她的敌人/捕食者解释:一旦被真菌感染,她便会面临痛苦的慢性死亡。毛虫知无不言,包括虫草这个不准确的命名的细节。她相信人类非常清楚虫草能为他们的健康和经济带来巨大的好处。她很现实地知道她在生命网络中的作用,也很谦逊地承认她和她的种群可以无私奉献、用于制药。她非常直接地进行对比:如果被渡鸦吃掉然后消化掉,她对这个世界几乎毫无用处,而真菌进入她身体转化成的复合体却可以被用作药材。毛虫正面回应了渡鸦的嘲弄。

艺术
The Arts

来写一首诗歌吧。挑选一个主题,给你的诗一个明确的定位,用象征和叙述表达你的想象。写诗是对外部世界和内心世界的观察。你的诗可以是关于任何主题或感觉。一只渡鸦(黑色的形状在蓝天中飞舞)……或一个生在毛虫身上的小蘑菇?写下第一行,然后让你的思绪任意漫游。想一想毛虫面临的威胁,或者渡鸦放弃的一顿午餐。想一想他们的感受。然后更具体一些,找出那些能够令你或你的读者为之惊喜的事情。

TEACHER AND PARENT GUIDE

思维拓展
Systems: Making the Connections

　　名字里含有什么？一种被称为草的真菌？一种实际上是毛虫的蚕？名字往往并不能准确表达事物本身，但即使我们没有使用准确的术语，我们也了解它们作为生命网络一环的重要性。毛虫一方面害怕被真菌感染，另一方面却对天然药材的重要性和自己对于生命网络的贡献充满赞赏。每个人都需要考虑他或她能为生活的改善做些什么——不仅仅是为今天或今年。

　　农民很早就认识到这一点，大多数农民都会保留两倍数量的种子，以防下个收获季收成不好。这样做就会有一段时期的缓冲来克服困难。现代生活越来越以个人为中心，很少有人愿意认清自己在生命网络中的角色，以及致力于人类共同利益的必要性。药物生产就是一个例子。更糟糕的是，我们所处的现代社会变得意识不到自身的脆弱性，成为一个消耗所有现有资源甚至是未来资源的社会。节约和恢复的概念在社会中越来越被忽视。

　　自然进程被生物制药工程所取代，我们为药物的大规模生产和更广泛普及感到高兴。然而我们好像忽略掉一个事实，即科技发展及其带来的进步令以虫草的持续性收获为生的农村社区的未来陷入危机。虽然这被称为进步，但我们往往没有料到，我们正在破坏喜马拉雅偏远山区存活了数千年的基本经济模式，使村庄人口减少，加速城市化进程，导致数以千计的人生活贫困。一旦我们意识到，作为生命网络的一部分，我们有一个远远大于个人利益的共同事业，我们就可以规划出能够满足当前所有人基本需求的社区，同时确保未来社会也这么做，甚至做得更好。

动手能力
Capacity to Implement

　　查阅喜马拉雅地区野生冬虫夏草的数量，并思考现有的野生冬虫夏草是否能满足当前世界的需求。现在检验一下两种做法：应用生物制药工程建立大规模生产，或者保留高原地区的采集传统及文化。接下来设计一个能让二者兼容的策略。与你的朋友及家人分享并讨论你的想法，告诉他们两者兼顾是可以实现的！

教师与家长指南

故事灵感来自

夏洛特·杜马雷斯·亨特
Charlotte Dumaresq Hunt

　　夏洛特·杜马雷斯·亨特生于1942年，被她的父亲昵称为"黛米"，因为她只有她妹妹的一半高。她在马萨诸塞州的剑桥长大，曾在墨西哥瓜纳华托、美国好莱坞和印度巴罗达大学学习，并获得富布莱特奖学金。她的丈夫黄泽思让她了解到中国的民俗、古代文化和历史。由她编著并配插图的儿童读物已经有300多本。她的作品《一粒大米：一个数学民间故事》介绍了数学的力量：一个乡村女孩要求国王在30天里每天给她的大米都是前一天的两倍，她用智慧战胜了国王，解救了饥饿的人。黛米因其作品多次获奖。她的故事由前美国第一夫人芭芭拉·布什选定，在美国国家广播电台ABC广播网节目"布什夫人的故事时间"上朗读。她的书已经被翻译成多种语言出版。

图书在版编目（CIP）数据

虫草：汉英对照/（比）冈特·鲍利著；（哥伦）凯瑟琳娜·巴赫绘；颜莹莹译．—上海：学林出版社，2017.10

（冈特生态童书．第四辑）

ISBN 978-7-5486-1240-7

Ⅰ．①虫… Ⅱ．①冈… ②凯… ③颜… Ⅲ．①生态环境－环境保护－儿童读物－汉、英 Ⅳ．① X171.1-49

中国版本图书馆 CIP 数据核字 (2017) 第 142689 号

© 2017 Gunter Pauli
著作权合同登记号　图字 09-2017-532 号

冈特生态童书
虫草

作　　者——	冈特·鲍利
译　　者——	颜莹莹
策　　划——	匡志强　张　蓉
责任编辑——	汤丹磊
装帧设计——	魏　来
出　　版——	上海世纪出版股份有限公司 学林出版社
	地　址：上海钦州南路 81 号　电话/传真：021-64515005
	网　址：www.xuelinpress.com
发　　行——	上海世纪出版股份有限公司发行中心
	（上海福建中路 193 号　网址：www.ewen.co）
印　　刷——	上海丽佳制版印刷有限公司
开　　本——	710×1020　1/16
印　　张——	2
字　　数——	5 万
版　　次——	2017 年 10 月第 1 版
	2017 年 10 月第 1 次印刷
书　　号——	ISBN 978-7-5486-1240-7/G.466
定　　价——	10.00 元

（如发生印刷、装订质量问题，读者可向工厂调换）